How To Browse The Internet With Less Data And Save Huge Money

By

Austin C. Christopher

Copyright

© 2020 Austin C. Christopher

All rights reserved

No part of this book may be reproduced, store in a retrieval device, or transmitted by any means, without the adequate consent and permission of the publisher.

Contact:

bornlandltd@gmail.com

Table of contents

Copyright *2*

Introduction *6*

HOW TO REDUCE YOUR EXPENSES ON INTERNET BROWSING *9*

Chapter 1 .... *18*

   DOWNLOAD VIDEOS BEFORE WATCHING..18

Chapter 2 .... *24*

   CHECK PEOPLE'S COMMENTS BEFORE DOWNLOADING OR WATCHING ANY VIDEO 24

Chapter 3 .... *27*

DOWNLOAD VIDEOS THAT SUITES THE RESOLUTION OF YOUR GADGET'S SCREEN . 27

Chapter 4 31

DON'T LEAVE YOUR DATA ON ALL THE TIME 31

Chapter 5 35

TURN OFF AUTO UPDATES 35

Chapter 6 39

DISABLE AUTO DOWNLOADS 39

Chapter 7 42

UNINSTALL THE APPS YOU ARE NOT USING 42

Chapter 8 45

DON'T BROWSE FROM YOUR CALL CREDIT 45

Chapter 9 .... *50*

BUY BULK DATA PLANS............................50

Chapter 10 .... *53*

REMOVE MALWARE AND VIRUS................53

CONCLUSION *57*

Introduction

This book was written to enlighten the public on how to browse the internet without wasting their money. The information contained in this book will teach you how to save a considerable amount of money while using the internet as you desire. Virtually everything in the world of today is going digital, and almost everyone browses the internet daily. The advancement of the internet makes information easily accessible, and digital transmission quite easy. This significant development has made all works of life and field of businesses to subscribe to digital means of transaction, interaction, and documentary. The concerned area of the use of the internet is the amount of money people spend per day, week, and month just to browse the web. It is quite excessive.

This book contains the tips and ideas you can apply while using the internet so you don't spend more than you should on an internet subscription. I advise you to gently read this book without any rush so you can be a beneficiary of the knowledge contained in this book. You will not waste your money anymore after reading this book.

HOW TO REDUCE YOUR EXPENSES ON INTERNET BROWSING

Everyone knows that we live in a computer age where virtually everything is going digital, and this has made the internet to grow so fast and also become so vast in the world. Some years ago, anyone who wants to attend church service must go to the place where the church building is physically built and join the worship. There was a time when you can only go to the physical market to buy or sell your goods. If you don't go to the market in those days, your products will remain with you. A few years ago, anyone who desires to go to school must go to where the school is physically located, get registered, and become a student. Every teaching must be conducted in the classroom. There were days when we could only watch movies and listen to the audio by inserting either the audio cd into the audio gadget or a radio cassette into a radio. There were days when we do many things offline. But today, everything has gone digital. These days, we have an online store, online schools, online churches, social media, computer applications, mobile applications, and many more. All these advancements are good because it has made life easier for us. You don't need to go to a post office before you can send a letter to someone. You can easily make use of your internet device to send messages and documents. You don't need to rent a shop before you can sell your goods. You can easily advertise your products online, and customers will order for it. You don't need to buy a CD or DVD before you watch movies. You can quickly go online and watch videos of your choice. These are amazing benefits of the internet. But there is something fundamental you need to consider and promptly amend so you don't spend more money than you should on the internet.

Improper use of the internet can make you waste the money you suppose to invest in a good business, buy a good house, buy a nice car, and do other essential things. The reason many people don't observe this is because they don't know the money equivalent of the time they spend and the data they use on the internet daily. These daily expenses may seem little to you, but if you check the accumulation of it in a month or year, you will know that you have spent a considerable amount of money.

This book is written to enlighten you and show you how to use the internet without wasting your money. Spending your cash carelessly will make you not to be as wealthy as you desire or supposed to. Many people want to be like certain wealthy people in the world without considering the secret of the constant increase in the wealth of such famous individuals. I have discovered that wealthy people do not waste their money. Instead, they invest their money wisely. That is why they keep increasing financially. Wealthy people ensure their profit covers their expenses so that they keep growing in wealth. But many poor people are more mindful of the pleasure they derive in browsing the internet than making profits. These have made the rich continue to be more affluent while the poor continue to be less privileged. If you genuinely want to be financially wealthy, then you must be careful about how you spend your money. You may not know this, but I have to tell you, "the internet is one of the major things that drains people's money" if it is not used carefully. Is it wrong to use the internet? The answer is, "No." But using the internet without being calculative will make you waste the money that supposes to earn you something meaningful in life. It will interest you to know that 70% of internet users are addicted to it, more especially, "the youths." If you doubt me look around your neighborhood or go to a busy place like Event Park and take a look at the people around. You will see that majority of the people are busy with their phones, iPads, or laptops. What are they doing, "They are browsing the internet."

Browsing the internet is very necessary. It helps us to access information quickly, but the point here is that many people ignorantly spend, or should I use the word "waste" their money all in the name of browsing. Have you wondered why your phone, iPad, laptop, and other internet-enabled devices have "Data Limit Settings"? The reason is that you can easily overuse the amount of data you plan or supposed to use per day without knowing. Therefore, these settings are programmed in your gadgets to help you limit your daily data usage. Unfortunately, many people do not make use of these settings either because they are not aware of it, they don't know how to use it, or they feel it's not necessary. But I have to tell you that these settings are required. If they are not, the manufacturers of these gadgets wouldn't have created it on the phone.

The knowledge you will derive from this book will amaze you and also help you to save a reasonable amount of money, which can be used for something meaningful. Your days of wasting money on the internet are over. After reading this book, you will gain the knowledge to spend what you suppose to spend and save what you suppose to keep. The following are the necessary things you need to do to save a reasonable amount of money while browsing the internet as you desire.

Chapter 1

DOWNLOAD VIDEOS BEFORE WATCHING

Streaming videos online is necessary if the video you are watching is live. But when you want to watch a video that is already uploaded online and it is available for downloads, it is necessary to download it into your gadget before watching the video so you can save some data, which is a value of money. Many people are not aware that streaming videos online cost more money and consume more data than downloading videos into the phone before watching. For example, if a one hour video you downloaded on your phone cost you 50mb, streaming the same one hour video online will cost you 150mb. Streaming videos online consumes over triple the data you use when you download videos before watching them.

The reason many people love to stream videos online is to save their phone memory space from being occupied. But I tell you that this reason is not good enough to waste your money. The best thing to do is to download the video, turn off your data after downloading the videos, and then watch the videos offline. After watching the video offline, delete the videos you don't want to keep on your phone so that your phone will still have space to contain other things. This is the wisdom you should apply.

You can search online for video download applications. There are many free video download apps that you can install on your phone, IPad, Laptop, and other internet-enabled gadgets. I download videos before I watch, and it's saving a lot of money for me. Stop streaming all videos online. Download videos, switch off your data, watch the video offline, and save money.

Note:

Please note that some websites and applications allow only video streaming on their websites. Other sites enable video download, but it must be directly from their official websites and not from a second or third party means. Some websites allow you to download their videos through other sites and applications. This simply means that every website has different policies, and you have to check the administration of the sites before downloading their contents. Please, don't violet the policy of any website.

Also, note that some of the contents on certain websites are copyrighted. This means that it cannot be edited and use for personal or commercial purposes without the permission of the owner. Many websites offer free non-copyrighted videos, audio, and images. Kindly go to those sites and make use of their contents as you wish. They are happy to provide free items to the public. I would have recommended some of the websites that offer free non-copyrighted materials so you can quickly go to their site, but many of them have trademark names that cannot be publicized without their consent. What I advise you to do is to go to a search engine on the internet and type "Free non-copyrighted videos, images, audio, etc.". You will see a good number of them to choose from. I hope this helps you. Please try not to violet the policy, rules, or regulation of any websites.

Chapter 2

CHECK PEOPLE'S COMMENTS BEFORE DOWNLOADING OR WATCHING ANY VIDEO

These days, many video creators and uploaders deceive people by writing titles that are different from the content of the video itself. They also use pictures that are different from the real content of the video just to get the attention of the public to watch their videos. After watching such a video, you get disappointed because what you expected based on the title and pictures is not what the video is all about. Any amount of data you spent while downloading or watching such a video is gone and wasted. Remember, data is money.

There are always comments below every video you see on social media and other platforms on the internet. Don't just see a video on the internet and begin to watch it. Check the comments below the video to know what people who have watched the video said about it. If the video is truly what the uploader claims it is, it will have good comments. Out of the ten comments below the video, seven will be positive. The positive comments mean that the video truly has good content, and you can download it. But if the video is not what the uploader claims it is, it will have negative comments. Out of the ten comments you see below the video, seven or eight will be negative. Don't waste your time downloading such a video because you will end up wasting your data and your money. Remember. Data is money. Check the comments below the video before you watch or download it.

Chapter 3

DOWNLOAD VIDEOS THAT SUITES THE RESOLUTION OF YOUR GADGET'S SCREEN

Many people waste money when they download videos online because they are not aware of this. Before I continue, I will like to let you know that choice of downloading videos based on its resolution is possible on the platforms, websites, and apps that are mostly programmed for video uploads and downloads. There are videos online that you don't have any other option than to download or watch it just as it is. But most of the sites where you can download videos and audios have these options that

I am talking about. If you want to download videos from most of the online video sharing platforms, the videos and audios have file sizes. These sizes do not determine how long the video will play but the size of the file. For example, if you want to download a one hour mp4 video, it will have different file sizes like; 1080P (720mb), 720P HD (420mb), 480P (260mb), 360P (220mb), 240P (105mb), and 144P (85mb). All these are the same one-hour videos but of different sizes. In a case like this, you don't need to download 1080P (720mb), which is the largest size and consume more megabytes and money. You can download the 480P (260mb), 360P (220mb), 240P (105mb), and 144P (85mb). Just download the one that would be clear on the screen of your phone or any gadget you want to watch it on. All you need is a version of a video that would be visually clear for your view.

The Audio files also have different sizes but at the same time duration. If you want to download a one-hour audio file, it will have different file sizes like; M4A 128 KBPS (60mb), Mp3 256 KBPS (120mb), and Mp3 128 KBPS (61mb). All these are the same one hour audio with different sizes. In a case like this, you don't need to down Mp3 256 KBPS

(120mb), which is the largest size and consumes more megabytes. You can download the M4A 128 KBPS (60mb), or Mp3 128 KBPS (61mb) version of the audio. All you need is audio that would sound clear to your hearing.

This knowledge will help you to use fewer data and save more money. Most times, I download the 360P version of the video, and it's always clear on my mobile phone and laptop, and it saves a tremendous amount of money for me. Stop wasting your hard-earned money. Use a little amount of data and save huge money.

Chapter 4

DON'T LEAVE YOUR DATA ON ALL THE TIME

Leaving your data on all the time will make you use much data and spend more money than you should. There are times you may want to leave your internet connection enabled for some crucial reasons. This is quite understandable. But when it is not necessary to leave your data on, please kingly turn it off.

One of the things that consume data is apps running on the background and some adverts that pops up from some apps as well. Many people are not away that apps run on the background of their phones and gadgets; that is why they often leave their data active. Most of these apps contain ads, and if the ads begin to pop up, it will be reducing your data without your awareness.

Some people leave their data on so they can receive notification of their email, chats, and other online updates. I am sure it is not every time that you need immediate notifications. You can go ahead and leave your data on when you need instant notifications. But when you are not expecting any urgent online message or information, please, kindly turn off your data so that you can use fewer data and save more money.

As a person, you know your activities both in business and other necessary meetings. You also know the time you will receive messages from your partners and clients. Why not take note of these times so that when it reaches, you can quickly turn on your data to activate the internet and receive notification of the messages you expect? This option would be a better idea instead of letting your data get drain-off for no reason. Many apps that are developed recently are advanced, and they have futures that pop up internet encoded software logically. So if you leave your data on, then you may lose massive data that could have served important purposes.

Henceforth, I want you to try this method, and it will amaze you to know that the data you consume in one month can serve you for two months. Remember, internet data is money. Save your data and save your money.

Chapter 5

TURN OFF AUTO UPDATES

Updates are essential for the applications on your phone and computer to work effectively. But you need to be mindful of what could happen when you leave your data updates active. Most of the official app store applications, update apps downloaded through them automatically. The negative side of it is that these applications update both the apps you use frequently and the apps you don't use at all, thereby wasting your data on something that is not useful to you. The best thing to do is to turn off the auto-update and update your desired apps manually so you can save your data and also keep your frequently used apps updated.

If you check some apps when it gives you the option to select either auto-updates or manual updates, it also recommends you set auto-update only when you are using Wi-Fi. Do you know why? It is because app updates consume data. One app update may not take much of your data, but the accumulation of data used to update five to ten apps will reduce your data drastically. Like I said earlier, it is vital to keep the apps on your phone and computer updated, but it has to be done manually to avoid wasting your internet data. Remember that data is money. Don't waste your money.

One good thing about turning off auto-update is that whenever there are updates available for any app on your phone or computer, it will notify you. Then you can manually update the apps that are useful to you. You can update your phone apps anytime you chose to do so. A manual update is the best.

Occasionally, one or two apps on your phone or computer may need their auto-updates to be turned on for specific reasons. In such a case, the app could be allowed to update automatically. But as for other apps that do not need an automatic update before they function accurately, I advise you to turn their auto-updates off so that it doesn't suck your internet data.

I encourage you to check your phone and computer right away and turn off auto-updates of the apps that do not need automatic updates to function. Save your data and save your money.

Chapter 6

DISABLE AUTO DOWNLOADS

Many apps have auto-download features for videos, audio, and images. Most social media and video download apps have these features, and they download videos, audio, and images automatically whenever you open the apps. If your data is on when you open the app, they start sucking immediately. It is necessary to disable the auto-download on such apps, so they don't waste your data unnecessarily. Remember, data is money.

Most times, when you go to your social media timelines and pages, we see lots of videos ranging from music, movies, sermon, and other kinds of videos. But the truth is that you will not watch all the videos because they may not be the kind of video you desire to view. If your data is on and the auto-download of the app you open is turned-on, it will start to download both the videos you like to watch and the one you don't like. These issues can be very annoying. It will end up wasting your time, money, and even your battery. The best thing to do is to turn off auto-download on your social medial and other chatting apps so that they don't download anything without your permission. When the auto-download is turned off, you can scroll through the videos, choose your desired videos, and download them. Doing this will save you some data and more money.

If the auto-download has been wasting your data, it will not waste it anymore because this knowledge will put an end to it. Just try it.

Chapter 7

UNINSTALL THE APPS YOU ARE NOT USING

Removing unnecessary apps from your phone does not only create space on your phone, it also saves you some data. Android phones and other internet-enabled mobile gadgets come with different kinds of apps, and some of these apps may not be useful to the owner of the phone even one to two years after purchasing the phone. This is not because the app is useless, but because the app may not be in line with the profession, lifestyle, academics, or business of the owner. For example, if a medical doctor buys a new phone and sees an app that speaks about how to make furniture, the app would be useless to the doctor because his profession as a medical doctor has nothing to do with furniture. If such an app remains on the doctor's phone, it will only occupy the phone space and also consume some data in the background. But if a professional furniture maker buys a phone that has this same app, it would be advantageous to him and instrumental to his work because the app contains information about his line of business. Therefore, he needs to keep the app.

This is exactly what I am talking about. Check the app on your phone and consider the purposes they serve. If there is an app on your phone that is not useful to you, kindly remove them so you can save your data. You can reinstall any app whenever you wish, so there is nothing to worry about it. Save your data and save your money.

Chapter 8

DON'T BROWSE FROM YOUR CALL CREDIT

Whenever you purchase a data plan, some network providers will ask you what they should do when your data finishes. Most of these network providers will display two options for you to choose from;

a) Do you want to continue to browse from your main call credit when your data finishes?

b) Do you want to stop browsing when your data finishes?

These are just two examples of the options which some of these network providers give to their subscribers when they purchase data. Some of the network providers offer more than two options. Let's consider the options above. If these two options are giving to you, the best option to choose is option "B," which says, "Do you want to stop browsing when your data finishes?". When this option is selected, your main call credit will not be affected when your data finishes because it has been programmed in the system of your network provider that your internet activities should not be charged from your main call credit. This wise decision will go a long way in saving you a good amount of money. But If you select option "A" which says; "Do you want to continue to browse from your main call credit when your data fishes?", your network provider will automatically charge your browsing activities from your main call credit when your data finishes because it has been programmed in their system that your internet activities should be charged automatically from your main call credit when your data finishes. This mistake will inevitably cost you a tremendous amount of money because browsing from your main call credit is about triple of data equivalent. For instance, if you spend $10 equivalent of data in one hour when you browse from your data, you will spend about $25 to $30 in one hour when you browse from your main call credit. The differences are always triple or even more.

Many people who are ignorant of this knowledge have lost an enormous amount of money because they browse from their main call credit when they run out of data. It is a good thing that you are reading this book because the knowledge you derive from this book will deliver you from being a victim of such significant losses. Stop browsing the internet from your call time, and you will save a reasonable amount of money that can be used for other essential things. Don't waste your money.

Chapter 9

BUY BULK DATA PLANS

Many people habitually like to buy daily and weekly data plans from their service providers. The reason for such subscriptions differs. Some of these people purchase daily and weekly data plans because they don't have the money to buy a monthly data plan. Whatsoever may be their reason, I want you to know that you spend more money when you purchase a daily and weekly data plan than when you buy a monthly data plan. It is like someone who went to a supermarket to buy a bottle of wine, and he was told that a bottle of wine is $10 while a park containing six bottles of the same wine is $45. If he buys the park containing six bottles of wine, he would be saving $15 that he can use to purchase other things he needed. I just use wine as an example. This method also applies to the data plans you buy from your internet service providers. You save money when you buy a monthly data plan than when you buy a daily or weekly data plan.

If you are among those who buy daily and weekly data plans, I want you to consider paying for a monthly subscription. If you don't have the money for a monthly subscription, it is quite understandable. But if you can afford to subscribe monthly, I advise you to do so, and it will amaze you how much money you will save for other important things you may need. Just try this, and you will see how it would be of great benefit to you.

Chapter 10

REMOVE MALWARE AND VIRUS

Malware and virus are another dangerous data drainer. They are so critical because they sneak into phones and computers without the consent of the owner. They do not come for good but to drain your data, steal your information, and destroy some files on your phone and computer. Most times, this malware installs apps on your phone without your awareness. Therefore, you must have a good antivirus on your phone and computer. Don't let your phone be vulnerable to the attack and menace of these viruses. Many phones and computers come with already installed antivirus. I advise you to check your phone and computers right away to know if there is antivirus in it. If you have an antivirus already installed on your phone, make sure it is turned on so it can protect your phone. But if you don't have it, kindly go online or a computer store and get a good antivirus for your phone and computer.

Make sure you scan your phone and computer manually at least once every week to ensure your gadget is virus-free. You can also set-up an auto-scan on the settings of your antivirus so it could scan your phone and computer automatically according to your settings.

Another thing you need to do is to make sure your phone does not allow the installation of an app from the unknown source without your permission. To get this done, follow these steps on your phone;

Go to settings, click on Security, and then uncheck "unknown resources."

When these settings are followed as directed, no app can be installed on your phone without your approval. This application will keep your phone safe.

Malware and virus perform many illicit activities on the phone and computers if the phone is not protected with a good antivirus. These illegitimate activities consumes lots of data. This menace cannot be underestimated nor overlooked. I encourage you to make sure that a good and functioning antivirus is installed on your phone and computer.

CONCLUSION

I believe the information contained in this book has enlightened you on how you can browse the internet as much as you like without wasting your precious money. People in the world are becoming wise enough to utilize their money by cutting down their expenses so they can save for other necessary things they desire to achieve in their lifetime. You should be wise too. You can browse the internet as often as possible, but don't let it empty your pocket at the end of the day. Apply the information contained in this book during your internet activities, and it will amaze you the amount of money you will save.

Thanks.

Contact:

bornlandltd@gmail.com

www.ingramcontent.com/pod-product-compliance
Lightning Source LLC
Chambersburg PA
CBHW031548210526
45464CB00003B/1209